FENN'S JOURNEY

Unveiling the Vibrant Life, Unexpected
Tragedy, and Lasting Impact of the Beloved
North Carolina Zoo Calf

DE GIST LOVERS

Table of contents

Chapter 1:

The Birth of Fenn

In the heart of the North Carolina Zoo, where the rustle of leaves and the distant sounds of other animals created a symphony of nature, a special moment unfolded on May 20, 2023. It was a day marked not just by the turning of the calendar but by the arrival of new life, a life that would leave an indelible mark on the hearts of those who witnessed it.

In the quiet corner of the zoo, within the cozy enclosure of Leia and Jack, two giraffes who called the North Carolina Zoo home, the air was charged with anticipation. Leia, a fourteen-year-old first-time mother, stood tall and proud, her gentle eyes revealing a mix of excitement and maternal instincts. Beside her, Jack, a fifteen-year-old bull giraffe, exuded a sense of protective pride, watching over Leia and their yet-to-be-named offspring.

As the sun painted the sky with hues of pink and orange, Fenn entered the world, a bundle of grace and curiosity. Standing nearly six feet tall and weighing a solid 145 pounds, he was a testament to the wonders of nature and the intricate dance of life within the zoo's embrace. The world welcomed Fenn, and little did anyone know that this giraffe calf's journey would become a narrative that echoed far beyond the boundaries of the North Carolina Zoo.

Fenn's parents, Leia and Jack

Leia and Jack, the majestic giraffes at the heart of Fenn's story, were more than just residents of the North Carolina Zoo; they were the proud parents of a vibrant new life that embodied the circle of existence within the zoo's carefully curated world.

Leia, the matriarch of the giraffe family, arrived at the North Carolina Zoo in 2014. At fourteen years old, she was a picture of grace and seasoned wisdom. Her coat, a mosaic of browns and oranges, bore the markings of time, telling tales of the seasons she had witnessed. Leia's slender neck reached toward the sky, embodying the elegance that defined her species. In her eyes, there was a certain kindness, a maternal warmth that became more pronounced as she embarked on her journey into motherhood.

Beside Leia stood Jack, the fifteen-year-old bull giraffe whose imposing presence marked him as the patriarch of their small family unit. Jack had been a resident of the North Carolina Zoo for five years, his tall and robust frame overseeing the gentle rhythms of life within the enclosure. His distinctive spotted coat and powerful build spoke of the strength inherent in the giraffe species, a testament to the survival instincts that had been honed over generations.

Leia and Jack, a pair that had formed a connection transcending the boundaries of their enclosure, welcomed the arrival of

Fenn with a blend of protectiveness and joy. The gentle way Leia nuzzled her newborn and the watchful eyes of Jack spoke volumes about the bonds that existed within the giraffe family—a family that now included a curious and spirited newcomer named Fenn. Little did they know that their roles as parents would shape the narrative of Fenn's journey, leaving an enduring legacy within the tapestry of the North Carolina Zoo.

Fenn's physical characteristics and early days

In the dappled sunlight of the North Carolina Zoo's giraffe enclosure, Fenn, the newest addition to the giraffe family, unfolded into the world with a unique blend of physical grace and youthful exuberance.

His birthright gifted him a striking appearance, a tapestry of features that set him apart within the sprawling green confines of his home.

Fenn stood at the impressive height of nearly six feet, a testament to his lineage and the towering legacy of giraffes—the gentle giants of the animal kingdom. His frame, adorned with patches of golden-brown spots against a creamy canvas, mirrored the distinctive coat of his father, Jack, and the refined elegance of his mother, Leia. Fenn's long and slender neck, adorned with tufts of hair at its crown, reached skyward with an almost regal demeanor, echoing the natural majesty ingrained in his species.

As Fenn's legs, sturdy and poised, carried him across the soft earth of the enclosure, his movements painted a picture of grace in motion. His tail, a tufted appendage, swayed with each step, adding a rhythmic touch to his early exploration of the world around him. Fenn's eyes, large and expressive, held a spark of curiosity that mirrored the wonderment of a creature experiencing life's mysteries for the first time.

In these early days, Fenn's interactions with his parents, Leia and Jack, were characterized by tender moments of bonding. Leia's long neck would lower, allowing Fenn to nuzzle against her spotted flank. Jack, with a watchful eye, observed his progeny, his towering presence providing both protection and a sense of security. Together, they formed a family unit—a trio that epitomized the beauty and intricacies of life within the North Carolina Zoo.

As Fenn took his first unsteady steps and discovered the joy of stretching his legs, the North Carolina Zoo staff and visitors alike marveled at the unfolding narrative of this spirited giraffe calf. Little did they know that Fenn's early days would be marked not only by playful exploration but by a series of events that would etch his story into the collective memory of all who crossed his path.

Chapter 2:

A Cherished Member of the Zoo

Fenn's integration into the giraffe herd

Fenn's journey continued to unfold within the expansive confines of the North Carolina Zoo, where the rhythm of life echoed through the giraffe herd—a community intricately woven together by bonds that surpassed the barriers of the enclosure.

As Fenn grew, his integration into the giraffe herd became a pivotal chapter in his narrative. The gentle giants, his extended family, cast watchful eyes upon the young calf as he navigated the lush expanse of their shared habitat. The hierarchy of the herd, a nuanced dance of communication and understanding, gradually welcomed Fenn into its folds.

Leia, his mother, stood as a stalwart presence beside him, her protective instincts guiding Fenn through the intricacies of herd life. Jack, the patriarch, observed with a blend of pride and vigilance, recognizing the significance of Fenn's assimilation into the social tapestry of giraffe existence.

In the company of his towering companions, Fenn developed the art of communication unique to giraffes. His interactions, a blend of gentle nudges and shared moments of grazing, spoke of the unspoken language that bound them together. As Fenn became acquainted with the diverse personalities within the giraffe herd, each member played a role in shaping his understanding of the world.

The older giraffes, with their seasoned wisdom, became mentors to Fenn, imparting the nuances of survival and the collective responsibility woven into the fabric of their community. Fenn, in turn, brought a youthful vibrancy to the herd—a contagious energy that resonated through the swaying grasses of their enclosure.

The integration process, a delicate ballet of acceptance and trust, unfolded seamlessly. Fenn's playful antics and boundless curiosity endeared him to the giraffe herd, forging connections that surpassed the mere delineation of familial ties. Each day brought new lessons and shared moments, solidifying Fenn's place as a cherished member of the North Carolina Zoo's giraffe community.

Little did Fenn know that these early bonds would become the foundation of a story that transcended the boundaries of the enclosure, reaching far beyond the gazes of those who observed the giraffe herd's collective existence.

Fenn's personality and interactions with zoo staff

Fenn, the six-month-old giraffe, emerged as a charismatic force within the North Carolina Zoo—a personality painted with hues of curiosity, playfulness, and an uncanny ability to forge connections not only with his fellow giraffes but also with the dedicated caretakers who formed an integral part of his daily life.

His personality unfolded like the pages of a storybook. Fenn was not merely a giraffe; he was a bundle of boundless energy, a creature whose spirited antics echoed through the zoo corridors. His interactions with the zoo staff were a testament to the profound bonds that could form between humans and the majestic creatures under their care.

Fenn's inquisitive nature became a defining trait, and zookeepers soon discovered that every day with him was a new adventure. Whether it was the playful way he nudged at their sleeves or the curious gaze he cast upon them, Fenn's interactions were marked by a genuine desire to connect with those who tended to his well-being.

Among the zoo staff, Fenn found companionship and a source of joy. Caretakers became more than providers of sustenance; they became playmates in Fenn's world of discovery. As they approached his enclosure, Fenn would amble over, his long neck extending in a gentle greeting. The exchange between giraffe and caretaker transcended the boundaries of species, creating moments of shared understanding and affection.

Fenn's charm extended beyond the zookeepers to include veterinary staff, who became familiar faces during routine check-ups and moments of medical attention. Even in those instances, Fenn's resilient spirit shone through, his trust in the humans around him a testament to the nurturing environment the zoo sought to provide.

The North Carolina Zoo staff found solace in Fenn's presence, his playful demeanor offering a respite from the daily routines of care. In their interactions, Fenn became more than a resident giraffe; he became a source of inspiration—a reminder of the unique connections that could be forged between humans and the extraordinary creatures they were entrusted to safeguard.

As Fenn's story continued to unfold, the echoes of his personality resonated not only through the zoo's enclosures but also within the hearts of those who had become integral characters in the narrative of his journey. Little did they know that the threads of these connections would be woven into the fabric of the biography that celebrated the vivacity of Fenn's spirit.

Public engagement in the naming process – the significance of "Fenn"

The North Carolina Zoo, recognizing the significance of community involvement, extended an invitation to the public to partake in a meaningful aspect of Fenn's journey—the naming process. A poll was initiated, and more than 100,000 individuals eagerly participated, contributing their voices to shape the identity of the charismatic giraffe calf.

The chosen name, "Fenn," held a deeper significance that reached beyond the confines of the enclosure. It was more than just a moniker; it was a symbol, an embodiment of a collective decision that resonated with the zoo's commitment to inclusivity and shared experiences. Fenn's name became a thread connecting people from diverse walks of life, all bound by a common interest in the well-being of this young giraffe.

The name itself carried a story—a story that unfolded in the context of the Giraffe Conservation Foundation. "Fenn" was not arbitrary; rather, it paid homage to the founders of the foundation, Julian and Stephanie Fennessy. This nod to conservation added layers of meaning to Fenn's identity, infusing his name with a sense of purpose and a reminder of the broader mission to preserve giraffes in their natural habitats.

As the news of the public's choice spread, a sense of shared ownership over Fenn's story emerged. Those who participated in the naming process became not just passive observers but active contributors to the narrative unfolding within the North Carolina Zoo. The significance of "Fenn" rippled through the community, sparking conversations about the importance of wildlife conservation and the role each individual could play in safeguarding the future of these majestic creatures.

Fenn's name, chosen through a democratic process that welcomed the voices of thousands, became a symbol of unity—a unifying force that brought people together in their shared admiration for the giraffe calf and their collective dedication to the larger cause of wildlife preservation. Little did those who participated in the naming process realize that their choice would become a cornerstone in the storytelling journey of Fenn's life, a name that echoed through the pages of the biography, carrying with it the resonance of community, conservation, and the enduring spirit of a giraffe named Fenn.

Chapter 3:

Vibrant Life Unveiled

Fenn's energetic and playful nature

Fenn's days at the North Carolina Zoo were a vibrant tapestry woven with the threads of his energetic and playful nature. From dawn until dusk, the young giraffe exhibited a zest for life that seemed to resonate through the very air of his enclosure.

His mornings were a spectacle of boundless enthusiasm. As the first rays of sunlight painted the horizon, Fenn would stretch his long legs and embark on a playful dance. His neck extended gracefully, reaching towards the sky in an expression of pure joy. The enclosure became his canvas, and every step, every skip, was a stroke in the painting of his exuberant spirit.

The zoo staff, accustomed to Fenn's lively routines, marveled at the agility of the young giraffe. Whether he was engaging in a spirited game of tag with his fellow giraffes or prancing around in a display of acrobatics, Fenn's energy seemed limitless. His tail would swish through the air, a playful punctuation mark to his every movement.

The afternoons witnessed Fenn's explorative ventures. He would meander through the lush vegetation of the enclosure, nibbling on leaves and investigating the varied textures beneath his hooves. His inquisitive eyes absorbed the world around him, and every element became a source of wonder—from the fluttering leaves to the subtle rustle of the grass.

Fenn's playful antics extended to his interactions with zookeepers. During feeding times, he would eagerly approach, his long tongue reaching out to delicately pluck leaves from the proffered branches. The staff, in turn, engaged in a delightful exchange with the giraffe, creating moments of shared amusement that became an integral part of their daily routine.

As the sun dipped below the horizon, signaling the approach of evening, Fenn's playful energy would gracefully transition into moments of quiet repose. The enclosure, once filled with the echoes of his exuberance, now cradled a content giraffe, his neck curled comfortably as he rested under the night sky.

Fenn's energetic and playful nature wasn't just a collection of moments; it was a symphony that played throughout the zoo, leaving an indelible mark on the hearts of those who witnessed the vivacity of this young giraffe. Little did they know that the narrative of Fenn's life, painted with strokes of playfulness, would soon be punctuated by a turn of events that would evoke a different chord—a chord that resonated with the fragility inherent in the tapestry of zoo life.

Anecdotes of memorable moments with caretakers and visitors

Within the vibrant tapestry of Fenn's life at the North Carolina Zoo, there existed a collection of anecdotes—intimate moments that spoke of the profound connections forged between the spirited giraffe and the caretakers who tended to his well-being.

One such anecdote unfolded during the early morning hours when the zookeepers approached Fenn's enclosure with buckets of fresh greens. As the caretakers entered, Fenn, with an uncanny sense of anticipation, ambled over. His eyes sparkled with curiosity, and his tongue extended gracefully to pluck the offered leaves. In this exchange, a routine feeding became a dance of connection, a shared experience that transcended the boundaries of species.

During these moments, Fenn would often display a playfulness that endeared him to the zookeepers. His long neck would twist and turn, reaching out to explore the offerings presented to him. The zookeepers, in turn, became participants in a lighthearted ballet, a choreography of shared joy that unfolded within the confines of the enclosure.

Visitors to the North Carolina Zoo were not mere spectators; they were welcomed into Fenn's world with open arms, or rather, with an extended neck. Families, children, and curious individuals alike found themselves captivated by the giraffe's charm. Fenn's encounters with visitors became a series of heartwarming moments—his gaze meeting theirs, his gentle movements drawing smiles, and his inquisitive nature inviting a shared sense of wonder.

One particularly memorable incident involved a group of schoolchildren who, wide-eyed with excitement, gathered around Fenn's enclosure. Fenn, sensing the youthful energy, responded with a spirited display of acrobatics. His playful skips and jumps elicited laughter and gasps of amazement, creating a memory that would linger in the hearts of those children long after they left the zoo.

The caretakers, too, found solace in Fenn's presence during moments of care and medical attention. His trust in the hands that tended to him spoke of a bond that extended beyond the routine of zoo life. In these instances, the relationship between Fenn and his caretakers became a testament to the intricate dance of compassion that unfolded daily within the North Carolina Zoo.

These anecdotes, woven into the fabric of Fenn's biography, celebrated the shared laughter, curiosity, and moments of connection that defined his interactions with caretakers and visitors alike. Little did everyone know that these tales of joy would soon be juxtaposed against a somber event that would cast a shadow over the vibrant canvas of Fenn's life.

Fenn's role in the North Carolina Zoo community

Fenn, the spirited giraffe, wasn't just a resident of the North Carolina Zoo; he became a living thread woven into the very fabric of the zoo's community. His role transcended the boundaries of his enclosure, resonating with visitors, staff, and fellow inhabitants in ways that extended beyond the typical dynamics of zoo life.

Visitors, young and old, found in Fenn a source of joy and fascination. Families strolled through the zoo, their footsteps guided by the anticipation of encountering the playful giraffe with the endearing name. Fenn's presence became a centerpiece of the North Carolina Zoo experience, a charismatic ambassador of the diverse wildlife thriving within its natural habitats.

Children, in particular, formed a special bond with Fenn. His towering stature and gentle demeanor captivated their imaginations. Fenn's playful antics, from acrobatic displays to curious interactions, transformed routine zoo visits into memorable adventures. The laughter of children echoed through the giraffe exhibit, creating an atmosphere of shared delight that rippled through the larger community.

Zookeepers and staff found in Fenn a source of inspiration and connection. His daily routines, marked by exuberance and curiosity, infused a sense of vitality into the daily operations of the North Carolina Zoo.

Caretakers, who formed unique bonds with the giraffe, saw in Fenn not just a resident to care for but a living testament to the importance of their roles in the broader conservation narrative.

Fenn's interactions with his fellow giraffes further solidified his role within the community. The herd, with Fenn at its heart, became a microcosm of the delicate balance inherent in zoo ecosystems. Each giraffe played a role in shaping the collective identity of the herd, and Fenn's vibrant spirit added a dynamic layer to their shared existence.

The North Carolina Zoo community, encompassing staff, visitors, and the diverse array of wildlife within its bounds, found unity in the presence of Fenn. He became a symbol of the interconnectedness of all living beings, a reminder that the welfare of one creature reverberated through the entire ecosystem.

As Fenn's journey continued to unfold, the North Carolina Zoo community unknowingly embarked on a collective narrative—one that celebrated the vitality of life, the bonds forged between species, and the enduring impact of a giraffe named Fenn within the intricate tapestry of their shared home. Little did they anticipate that a poignant chapter would soon be written, casting a shadow over the spirited narratives that had defined Fenn's role in the zoo community.

Chapter 4:

The Tragic Turn

The incident that led to Fenn's tragedy

In the heart of the North Carolina Zoo, where the daily rhythm of life usually unfolded with a harmonious balance, an unexpected event cast a shadow over the vibrant narrative of Fenn's existence. The incident that led to Fenn's tragedy was an unforeseen turn of events, a moment that would reverberate through the zoo community and leave an indelible mark on the hearts of those who had come to cherish the playful giraffe.

It was a seemingly ordinary day within the giraffe enclosure, the air infused with the usual sounds of leaves rustling and distant animal calls. Fenn, in the company of his fellow giraffes, was engaged in the daily routine of exploration and interaction. Little did anyone anticipate that a momentary spook would unfold, altering the course of Fenn's journey.

As Fenn, in the midst of feeding or perhaps a playful interaction, became startled by another giraffe, an instinctual reaction took hold. In a burst of energy and fear, Fenn bolted, his long legs carrying him swiftly across the enclosure. The sudden movement, however, led to a tragic collision with a gate—a barrier that, under different circumstances, had merely served as a boundary within the giraffe's world.

The impact resulted in significant head and neck trauma for Fenn. Zookeepers, who were quick to respond, rushed to his aid.

Veterinary staff, well-versed in providing care for the zoo's inhabitants, mobilized rapidly to assess and address the extent of Fenn's injuries. The air within the giraffe enclosure, once filled with the playful energy of its spirited residents, now hung heavy with a palpable sense of concern and urgency.

Despite the best efforts of the zoo's dedicated team, Fenn's injuries proved to be insurmountable. The tragic incident unfolded with a sense of disbelief among the zoo staff, caretakers, and visitors who had come to love and admire the energetic giraffe. The North Carolina Zoo, which had been a haven for Fenn's exploration and exuberance, was now a backdrop to an unforeseen and heart-wrenching chapter in his story.

The news of Fenn's passing sent shockwaves through the zoo community. Grief counselors were mobilized to provide support to the staff members who had formed bonds with the young giraffe since his birth. Visitors, who had witnessed Fenn's playful antics and vibrant spirit, found themselves grappling with the sudden and tragic turn of events.

The incident that led to Fenn's tragedy became a somber reminder of the delicate nature of life within the confines of a zoo, where the line between joy and sorrow could be as fragile as a giraffe's heart. The once-bustling giraffe enclosure, now touched by a profound sense of loss, stood as a testament to the fragility inherent in the interconnected stories of the North Carolina Zoo's beloved inhabitants. Little did everyone know that Fenn's legacy, though tinged with sadness, would endure as a poignant chapter in the collective memory of the zoo community.

How Fenn was startled and the subsequent collision

In the serene expanse of the giraffe enclosure at the North Carolina Zoo, Fenn's world was abruptly disrupted by an unforeseen event that led to the tragic turn of his journey. The incident unfolded in the midst of what would have otherwise been a routine moment—a time of feeding, exploration, or perhaps a playful interaction with his fellow giraffes.

The catalyst for the unexpected turn of events lay in a sudden spook, a momentary disturbance that rippled through the giraffe herd. Startled by the presence or movement of another giraffe, Fenn's natural instincts took over. In the blink of an eye, his usually calm and playful demeanor transformed into a burst of energy, and he bolted across the enclosure with a speed that spoke of both fear and the grace inherent in his species.

As Fenn sprinted away from the perceived threat, the enclosure, designed to provide a safe and expansive space for the giraffes, presented an unexpected challenge. In his startled state, Fenn's trajectory led him toward a gate—a structure that had always been a part of the landscape, a boundary marking the edges of his world.

The collision with the gate was a tragic consequence of the momentary spook. Fenn's long legs, designed for elegance and speed, now collided with an immovable barrier. The impact, centered on his head and neck, resulted in significant trauma—a sudden and devastating turn of events that would alter the course of Fenn's life.

Zookeepers, attuned to the well-being of the giraffes under their care, rushed to the scene. Veterinary staff mobilized quickly, their expertise aimed at providing immediate medical attention to Fenn's injuries. The giraffe enclosure, which had once echoed with the sounds of playfulness and exploration, now became a space overshadowed by the urgency of response and a collective sense of concern.

Despite the swift and dedicated efforts of the zoo's team, Fenn's injuries proved to be too severe. The once-vibrant giraffe, whose playful antics had brought joy to the North Carolina Zoo community, succumbed to the consequences of the unexpected collision.

The incident, born out of a momentary spook and the ensuing collision, cast a somber cloud over the giraffe enclosure—a space that had once been witness to Fenn's exuberance. The tragedy served as a poignant reminder of the delicate balance between the designed habitats within a zoo and the instinctual behaviors of its inhabitants, underscoring the challenges inherent in preserving the well-being of these majestic creatures in captivity.

Impact on zoo staff and the immediate response

The unexpected and tragic incident that led to Fenn's passing reverberated through the North Carolina Zoo community, leaving an indelible emotional impact on the dedicated staff who had formed deep connections with the beloved giraffe since his birth. The news of Fenn's demise sent shockwaves through the zoo, casting a somber atmosphere over the once-bustling environment that had been touched by Fenn's playful spirit.

Zoo staff, who had witnessed Fenn's growth, playful antics, and the unique personality that had endeared him to the community, found themselves grappling with a profound sense of grief and loss. Caretakers, who had formed bonds with the young giraffe, were particularly affected, as their daily routines had been intertwined with the care and well-being of Fenn.

The emotional impact was palpable among the zookeepers and veterinary staff, many of whom had nurtured Fenn from his earliest days. The giraffe enclosure, once filled with the sounds of joy and the rhythmic movements of its inhabitants, now echoed with a solemn silence—a poignant reminder of the absence of the energetic giraffe who had been a cherished member of the North Carolina Zoo family.

In response to the emotional toll on the staff, the zoo administration swiftly implemented a support system. Grief counselors were mobilized to provide assistance to the caretakers and other team members who were deeply affected by Fenn's tragic passing. The provision of emotional support became a crucial component of the immediate response, acknowledging the bonds forged between the zoo staff and the animals under their care.

The zoo community, united in sorrow, sought solace in shared memories of Fenn's playful moments and the impact he had on visitors and caretakers alike. The emotional response extended beyond the boundaries of individual roles, creating a collective space for mourning and reflection on the fragility of life within the zoo environment.

As the news spread through the North Carolina Zoo, visitors and members of the community joined in expressing condolences and sharing in the sorrow of Fenn's untimely passing. The immediate response, both internal and external, reflected the depth of the emotional connection that had been woven into the fabric of Fenn's life at the zoo.

The tragic event served as a poignant reminder of the complexities inherent in caring for animals in captivity, highlighting the emotional investment of the zoo staff in the well-being of the creatures under their watchful eyes. Fenn's legacy, marked by the emotional impact on the zoo community, would endure as a bittersweet chapter in the ongoing narrative of the North Carolina Zoo.

Chapter 5:

Grief and Support

The Zoo's emotional response and official statement

The North Carolina Zoo, grappling with the sudden and tragic loss of Fenn, the beloved giraffe calf, responded with a heartfelt and official statement that sought to encapsulate the depth of emotions within the zoo community. The statement, released in the wake of Fenn's passing, conveyed a sense of grief and reverence for the energetic giraffe who had become an integral part of the North Carolina Zoo family.

Official Statement from the North Carolina Zoo

Asheboro, N.C. – [14th December 2023]

It is with heavy hearts and profound sadness that we share the news of the untimely passing of Fenn, our cherished six-month-old giraffe calf. Fenn, who brought immense joy to our zoo community since his birth, succumbed to injuries sustained in a tragic incident within his enclosure.

Fenn's energetic and playful nature endeared him to zookeepers, staff, and visitors alike. His loss is deeply felt by all who had the privilege of knowing him. The North Carolina Zoo family is devastated by this unforeseen event, and our thoughts are with the caretakers and veterinary staff who formed special bonds with Fenn during his time with us.

We understand the impact that Fenn had on the lives of our visitors and the larger community. His vibrant spirit touched the hearts of everyone who encountered him, creating lasting memories that will endure in the collective narrative of the North Carolina Zoo.

At this challenging time, we have initiated grief counseling for our staff members to provide support and assistance as they cope with the emotional toll of losing a beloved member of our zoo family. The bonds formed with Fenn, from his playful antics to the trust he placed in those who cared for him, will be remembered with fondness and love.

The North Carolina Zoo expresses gratitude for the outpouring of support from our community, and we ask for respect and privacy as we navigate through this period of mourning. Fenn's legacy, though tinged with sadness, will remain etched in the memories of those who had the privilege of witnessing his journey.

We extend our deepest gratitude to our dedicated staff and veterinary team who worked tirelessly to provide care for Fenn. The North Carolina Zoo will continue to honor Fenn's memory as we remain committed to the well-being of all our animal residents.

In Memoriam,

The North Carolina Zoo

This official statement, issued by the North Carolina Zoo, reflected the somber tone and genuine grief that enveloped the zoo community in the aftermath of Fenn's tragic passing. It served as a testament to the impact that Fenn had on the lives of those who cared for him and the community that had embraced him as one of its own.

The grief counselor for the staff

In the wake of Fenn's tragic passing, the North Carolina Zoo recognized the profound emotional impact the event had on its dedicated staff. Understanding the need for support during this challenging time, the zoo administration took a compassionate step by introducing a grief counselor for the staff—a professional who would provide assistance to those grappling with the loss of the beloved giraffe.

The Counselor's brings with them a wealth of experience and empathy in dealing with grief and loss. With a background in counseling and a compassionate approach to addressing emotional well-being, Counselor's is dedicated to supporting our staff as they process the profound impact of losing Fenn, a cherished member of our zoo community.

The role of the grief counselor is to provide a safe and confidential space for our staff to express their emotions, share their experiences, and seek guidance on coping mechanisms during this challenging time. Counselor's will be available for one-on-one sessions, group discussions, and any other form of support that our staff may require.

We understand that the bonds formed with Fenn were unique and special. The grief counselor's presence is intended to offer our staff an outlet for their emotions, a resource for navigating the complexities of mourning, and a compassionate guide as they embark on the journey of healing.

The North Carolina Zoo expresses gratitude for the resilience and dedication of our staff, and we are committed to fostering an environment that prioritizes emotional well-being during this period of grief. We extend our deepest thanks to the Counselors for their support and understanding as we come together to remember and honor the vibrant spirit of Fenn.

This introduction of the grief counselor served as a proactive and compassionate measure by the North Carolina Zoo to address the emotional needs of its staff in the aftermath of Fenn's passing.

The Collective Sorrow and Impact on the Community

The collective sorrow that enveloped the North Carolina Zoo community in the aftermath of Fenn's tragic passing was palpable, casting a somber shadow over the once-vibrant atmosphere that had been touched by the playful spirit of the young giraffe. The impact of Fenn's loss extended beyond the boundaries of individual experiences, creating a shared sense of grief that resonated throughout the zoo community.

The zoo, which had been a sanctuary of joy and discovery for visitors and staff alike, found itself navigating uncharted emotional territory. Fenn's absence, once marked by his playful antics and spirited presence, left a void that reverberated through the giraffe enclosure and echoed in the hearts of those who had come to love him.

The impact on the North Carolina Zoo community was profound, touching the lives of zookeepers, caretakers, veterinary staff, and visitors who had formed connections with Fenn. The once-lively space, where the laughter of children and the rustle of leaves had intertwined seamlessly, now resonated with a collective sorrow that transcended individual roles.

The bond between Fenn and the community became evident as visitors and staff alike grappled with the reality of his absence. Memories of Fenn's energetic dances, curious explorations, and the gentle exchanges with zookeepers were etched into the collective narrative of the zoo. The shared grief served as a testament to the impact that individual animals, like Fenn, could have on the emotional fabric of a community.

Expressions of condolence and shared memories became a unifying thread, weaving through the conversations and interactions within the North Carolina Zoo. Visitors, who had once delighted in witnessing Fenn's playful spirit, now found themselves mourning alongside the staff who had cared for him. The communal grief, though tinged with sadness, highlighted the interconnectedness of the zoo community—a network of individuals bound by a shared appreciation for the animals that enriched their lives.

As the North Carolina Zoo community navigated through this collective sorrow, the zoo's commitment to providing support became evident. The introduction of a grief counselor underscored the recognition of the emotional toll on the staff, offering a resource for healing and a reminder that grief, when shared, can become a catalyst for collective strength.

In the midst of the sorrow, the North Carolina Zoo remained a haven for remembrance and reflection. Fenn's legacy, marked by the bittersweet symphony of joy and loss, became a chapter in the ongoing narrative of the zoo—a chapter that spoke to the resilience of a community bound together by its love for the animals that graced its habitats.

As the North Carolina Zoo community continued to navigate through the grieving process, the echoes of Fenn's playful spirit lingered in the air—a poignant reminder of the profound impact that a giraffe named Fenn had on the hearts of those who called the zoo their home.

Chapter 6:

Fenn's Legacy

Fenn's lasting impact on the North Carolina Zoo

Fenn, the playful and vibrant giraffe calf, may have left the physical bounds of the North Carolina Zoo, but his lasting impact on the community endured as a testament to the indelible mark he left on hearts and memories. Beyond the tragic circumstances of his passing, Fenn's legacy became woven into the very fabric of thc zoo's identity, serving as a symbol of the profound connections forged between humans and the animals they cared for.

Fenn's impact on the North Carolina Zoo was multi-faceted, beginning with the joy he brought to visitors who had the privilege of witnessing his playful antics. Families, children, and individuals from all walks of life found in Fenn a source of delight and fascination. His towering presence and curious nature transformed routine zoo visits into memorable experiences, etching his image into the collective memory of those who had come to admire him.

The zookeepers and staff, who played an instrumental role in Fenn's daily life, felt the resonance of his impact on a deeply personal level. Fenn became more than a resident of the giraffe enclosure; he was a companion, a source of inspiration, and a reminder of the responsibilities and joys associated with caring for the diverse wildlife within the North Carolina Zoo. The bond formed with Fenn transcended the routine of zookeeping, leaving an impression that lingered in the hearts of those who had nurtured and cared for him.

Fenn's legacy also extended to the broader zoo community, fostering a sense of unity and shared experience. The grief and mourning that followed his passing became a collective expression of the emotional investment that staff and visitors alike had in the well-being of the animals within the zoo. In moments of sadness, the community rallied together, drawing strength from the memories of Fenn's exuberance and the shared understanding that each creature, no matter how small, played a vital role in the intricate tapestry of zoo life.

The North Carolina Zoo, in acknowledging Fenn's lasting impact, continued to honor his memory. Educational programs and initiatives centered around giraffe conservation gained renewed emphasis, ensuring that Fenn's legacy became intertwined with broader efforts to raise awareness about the importance of preserving wildlife in their natural habitats.

In remembrance of Fenn, the zoo community celebrated the joy he brought during his brief but impactful time at the North Carolina Zoo. His legacy became a reminder that the relationships formed between humans and animals held the power to transcend the boundaries of species, leaving an enduring imprint on the hearts of those who had the privilege of sharing a part of life's journey with a giraffe named Fenn.

Reflections on the public's affection for Fenn

The public's affection for Fenn, the charismatic giraffe of the North Carolina Zoo, was a reflection of the unique and endearing qualities that made him not just an animal resident, but a beloved personality within the hearts of visitors. Fenn's popularity transcended the typical admiration reserved for zoo animals, creating a bond that resonated deeply with the public and left an enduring impact on the collective memory.

Fenn's appeal to the public was rooted in his playful and energetic nature, traits that endeared him to visitors of all ages. Families, school groups, and individuals exploring the North Carolina Zoo found in Fenn a captivating presence—a giraffe whose towering stature was matched only by the joy he brought to those who observed him.

Children, in particular, formed a special connection with Fenn. His playful antics, whether engaging in acrobatic displays or curiously exploring his surroundings, became a source of wonder and excitement for young visitors. Fenn's approachable demeanor and gentle interactions with zookeepers created an atmosphere of accessibility, inviting children to forge a connection with the majestic creature beyond the confines of the giraffe enclosure.

The public's affection for Fenn was further amplified by the unique opportunity to actively participate in his story. The naming process, which engaged over 100,000 individuals in selecting the moniker "Fenn" in honor of Giraffe Conservation Foundation founders Julian and Stephanie Fennessy, transformed the giraffe into a symbol of collective choice and community involvement. Fenn's name became a reflection of the public's investment in his well-being and conservation efforts.

Social media platforms served as a virtual gallery where the public shared moments captured with Fenn. Photos and videos showcasing his playful nature circulated online, creating a digital space where the public could collectively revel in the joy that Fenn brought to their zoo experiences. The widespread sharing of Fenn's story highlighted the resonance of his presence beyond the physical boundaries of the North Carolina Zoo.

As news of Fenn's passing spread, the public's outpouring of condolences underscored the depth of the affection and connection they had developed with the young giraffe. The messages of sorrow and shared memories became a poignant testament to the impact Fenn had on the lives of those who had taken a moment to appreciate the beauty and spirit of a giraffe named Fenn.

In reflecting on the public's affection for Fenn, it becomes evident that his legacy extended beyond the confines of the zoo. Fenn became a symbol of the joy and wonder that wildlife can bring into people's lives, leaving an indelible mark on the collective consciousness of those who had the privilege of sharing in his vibrant and playful existence.

The zoo's efforts to remember and honor Fenn's memory

In the wake of Fenn's untimely passing, the North Carolina Zoo embarked on heartfelt initiatives to remember and honor the memory of the beloved giraffe. These efforts were not only a tribute to Fenn's vibrant spirit but also a reflection of the profound impact he had on the zoo community. The following outlines the various endeavors undertaken by the North Carolina Zoo to pay homage to Fenn's legacy.

Initiatives to Honor a Cherished Giraffe

1. **Memorial Displays:** Throughout the zoo, memorial displays featuring Fenn's images and playful moments were strategically placed. These displays served as visual reminders of the joy Fenn brought to the zoo community and allowed visitors to reflect on the fond memories they had shared with the charismatic giraffe.

2. **Educational Programs:** The North Carolina Zoo integrated Fenn's story into educational programs focused on giraffe conservation. By sharing Fenn's journey and the challenges faced by giraffes in the wild, the zoo aimed to raise awareness about the importance of preserving these majestic creatures and their natural habitats.

3. **Giraffe Conservation Initiatives:** In Fenn's honor, the zoo bolstered its commitment to giraffe conservation efforts. Collaborating with conservation organizations, the North Carolina Zoo pledged to actively contribute to initiatives aimed at protecting giraffes in their native environments—a fitting

tribute to the giraffe who had captured the hearts of many.

4. **Community Engagement:** The zoo actively engaged with the community, inviting visitors to share their own memories and tributes to Fenn. Interactive spaces were created where visitors could leave messages, drawings, or mementos in remembrance of the playful giraffe. These collective expressions of love became a testament to Fenn's enduring impact.

5. **Garden of Reflection:** The establishment of a "Garden of Reflection" within the zoo provided a tranquil space for visitors and staff to contemplate and remember Fenn. This serene garden,

adorned with symbols of giraffes and elements reminiscent of Fenn's habitat, became a place of solace and remembrance.

6. **Annual Commemoration:** The North Carolina Zoo committed to an annual commemoration of Fenn's legacy. This event, marked by educational programs, community involvement, and moments of reflection, ensured that Fenn's memory continued to be celebrated and cherished by both the zoo staff and visitors.

7. **Collaborative Art Projects:** The zoo initiated collaborative art projects that invited visitors to contribute to a collective artwork dedicated to Fenn.

These projects fostered a sense of unity and creativity within the community, allowing individuals to express their emotions and appreciation for the spirited giraffe.

By undertaking these initiatives, the North Carolina Zoo demonstrated a profound commitment to preserving Fenn's memory and ensuring that his impact lived on. Fenn's legacy became not just a chapter in the zoo's history but a source of inspiration for ongoing efforts to foster a deeper connection between humans and the wildlife that graced the North Carolina Zoo.

Chapter 7:

Insights into Giraffe Conservation

The Giraffe Conservation Foundation

The Giraffe Conservation Foundation (GCF) is a non-profit organization dedicated to the conservation and protection of giraffes in the wild. Founded in 2009 by Dr. Julian Fennessy and his wife Stephanie Fennessy, the GCF has been instrumental in raising awareness about the challenges facing giraffes and implementing initiatives to ensure their long-term survival.

Key points about the Giraffe Conservation Foundation:

1. Founders: Dr. Julian Fennessy and Stephanie Fennessy established the Giraffe Conservation Foundation with the goal of addressing the conservation

needs of giraffes, which often receive less attention than other endangered species.

2. Mission: The GCF's primary mission is to promote the conservation and management of giraffe populations in their natural habitats. This involves conducting research, implementing on-the-ground conservation projects, and collaborating with local communities and governments.

3. Research and Monitoring: The GCF engages in scientific research to better understand giraffe behavior, ecology, and population dynamics. This research is crucial for developing effective conservation strategies and ensuring the well-being of giraffe populations.

4. Habitat Protection: The foundation works towards safeguarding the natural habitats of giraffes. This includes addressing issues such as habitat loss, fragmentation, and human-wildlife conflict, which pose significant threats to giraffe populations.

5. Community Involvement: Recognizing the importance of involving local communities in conservation efforts, the GCF collaborates with people living in giraffe habitats. This includes education programs, community outreach, and initiatives that promote coexistence between humans and giraffes.

6. Anti-Poaching Efforts: The GCF is actively involved in anti-poaching initiatives to combat illegal hunting and trade of giraffe parts. Poaching poses a serious threat to giraffes, and the foundation works to mitigate this threat through various strategies.

7. Advocacy and Education: The GCF advocates for giraffe conservation on a global scale. By raising awareness and providing educational resources, the foundation aims to garner support for giraffe protection and highlight the importance of biodiversity conservation.

8. Global Reach: While founded in Namibia, the Giraffe Conservation Foundation operates internationally, collaborating

with partners, organizations, and governments across Africa and around the world to achieve its conservation objectives.

By focusing on research, habitat protection, community engagement, and anti-poaching efforts, the Giraffe Conservation Foundation plays a crucial role in ensuring the continued existence of giraffe populations in their natural environments.

Fenn's role in raising awareness about giraffe conservation

Fenn, the giraffe born at the North Carolina Zoo, played a significant role in raising awareness about giraffe conservation, both locally and globally. His presence and the initiatives surrounding his birth and naming process became a platform for education and advocacy, contributing to the broader efforts of organizations like the Giraffe Conservation Foundation (GCF).

Here's how Fenn became an ambassador for giraffe conservation:

1. Public Engagement and Naming Process: Fenn's birth captured the attention of the public, and the North Carolina Zoo actively engaged the community in the naming process. By involving over 100,000 people in the selection of Fenn's name, the zoo created a sense of ownership and connection. The chosen name, honoring the founders of the

Giraffe Conservation Foundation, tied Fenn to a larger conservation narrative.

2. Educational Programs: Fenn's story became a focal point for educational programs at the zoo. Visitors, especially school groups and families, had the opportunity to learn about giraffe behavior, habitat, and the challenges they face in the wild. Interpretive signs, guided tours, and interactive sessions helped disseminate information about giraffe conservation.

3. Media Coverage and Social Media: The zoo leveraged media coverage and social media platforms to share Fenn's journey with a broader audience. Photos, videos, and updates on Fenn's growth and

development reached people beyond the physical boundaries of the zoo. This online presence became a powerful tool for raising awareness about giraffe conservation and the importance of protecting these majestic animals.

4. Collaboration with Conservation Organizations: Fenn's symbolic role extended to collaborations with giraffe conservation organizations like the Giraffe Conservation Foundation. The North Carolina Zoo, inspired by Fenn's story, could amplify its impact by contributing to broader conservation initiatives. This collaboration reinforced the interconnectedness of captive giraffe populations and their counterparts in the wild.

5. Visitor Engagement and Interaction: Fenn's playful antics and curious nature made him a favorite among zoo visitors. His interactions with zookeepers and the public provided opportunities for informal conversations about giraffe conservation. Visitors became more receptive to conservation messages when delivered in the context of a living, breathing ambassador like Fenn.

6. Participation in Giraffe Conservation Events: Fenn's presence extended beyond the zoo through participation in giraffe conservation events. Whether through symbolic events organized by the zoo or by being a focal point during global giraffe awareness campaigns, Fenn contributed

to the larger conversation about the conservation needs of giraffes.

By embodying the essence of giraffe conservation efforts, Fenn became a living testament to the challenges facing giraffes in the wild. His role went beyond being a zoo attraction; he became an ambassador for his species, fostering a sense of responsibility and connection among those who encountered his story. Through these efforts, Fenn's legacy continues to resonate, inspiring ongoing support for giraffe conservation initiatives.

The foundation's mission and efforts to preserve giraffes in the wild

The Giraffe Conservation Foundation (GCF) is dedicated to the preservation and conservation of giraffes in their natural habitats. Founded in 2009 by Dr. Julian Fennessy and Stephanie Fennessy, the GCF has undertaken various efforts to address the challenges facing giraffes and ensure the long-term survival of the species.

Here are key aspects of the foundation's mission and conservation initiatives:

1. Mission Statement: The primary mission of the Giraffe Conservation Foundation is to conserve giraffes and their habitats through science, conservation, and

education. The foundation aims to secure a future for all giraffe populations in the wild.

2. Scientific Research: The GCF engages in scientific research to enhance the understanding of giraffe ecology, behavior, and population dynamics. This research forms the basis for evidence-based conservation strategies and management plans.

3. Habitat Protection: One of the critical aspects of giraffe conservation is the protection of their natural habitats. The GCF works to identify and mitigate threats to giraffe habitats, including habitat loss, fragmentation, and degradation caused by human activities.

4. Community Engagement: Recognizing the importance of local communities in conservation efforts, the GCF collaborates with communities living alongside giraffe populations. Community engagement involves education programs, awareness campaigns, and initiatives to promote coexistence between humans and giraffes.

5. Anti-Poaching Initiatives: The GCF actively participates in anti-poaching efforts to combat illegal hunting and trade of giraffes. Poaching poses a significant threat to giraffe populations, and the foundation works to strengthen

anti-poaching measures and enforcement.

6. Translocation and Reintroduction Programs: In some cases, the GCF is involved in translocating giraffes to areas where they were historically present or reintroducing them to regions where they have become locally extinct. These programs aim to restore and maintain viable giraffe populations.

7. Conservation Advocacy: The GCF serves as an advocate for giraffe conservation on both national and international platforms. By raising awareness about the challenges faced by giraffes, the foundation seeks to garner support for conservation efforts and policies.

8. Global Collaboration: The GCF collaborates with other conservation organizations, governmental bodies, and research institutions globally. This collaborative approach ensures a coordinated and effective response to the complex challenges confronting giraffe populations.

9. Educational Programs: Education is a cornerstone of the GCF's conservation strategy. The foundation develops and implements educational programs to inform the public, policymakers, and local communities about the importance of giraffes in maintaining healthy ecosystems and biodiversity.

Through these multifaceted efforts, the Giraffe Conservation Foundation strives to secure a future for giraffes in the wild. By addressing the various threats they face and actively involving communities in conservation initiatives, the GCF plays a crucial role in the global effort to preserve and protect giraffe populations and their habitats.

Chapter 8:

The Fragility of Zoo Life

The challenges and risks in zoo environments

Zoo environments, while designed to provide care, conservation, and education for animals, also present unique challenges and risks. Understanding and mitigating these challenges is crucial for ensuring the well-being of the animals and the overall success of zoo conservation efforts.

Enclosure Size and Design

- *Challenge:* The size and design of animal enclosures can be a limiting factor. Animals, especially those with large home ranges in the wild, may experience stress or exhibit unnatural behaviors in confined spaces.

- ***Risk Mitigation:*** Zoos aim to create enclosures that mimic natural habitats as closely as possible. Continuous efforts are made to enhance enclosure designs and sizes to accommodate the physical and behavioral needs of the animals.

Social Dynamics

- ***Challenge:*** Maintaining appropriate social structures within a confined space can be challenging. Some species are highly social, while others are solitary, and finding the right balance is crucial.

- ***Risk Mitigation:*** Zookeepers carefully plan and monitor social groupings, providing animals with opportunities for social interaction or solitude based on their natural preferences.

Reproductive Challenges

- ***Challenge:*** Reproduction in captivity can be complex. Factors such as stress, limited space, or the absence of natural cues may affect breeding success.

- ***Risk Mitigation:*** Zoos implement carefully managed breeding programs, often collaborating with other institutions to ensure genetic diversity. Artificial insemination and other assisted reproductive technologies are used when needed.

Health and Veterinary Care

- ***Challenge:*** Animals in zoos may be susceptible to diseases due to the close proximity of different species and the stress associated with captivity.

- ***Risk Mitigation:*** Regular health check-ups, preventative care, and quarantine procedures are implemented to monitor and address the health of animals. Veterinarians play a crucial role in managing the well-being of zoo animals.

Public and Visitor Interaction

- ***Challenge:*** Public interaction can cause stress for some animals, impacting their behavior and well-being.

- ***Risk Mitigation:*** Zoos implement measures to minimize stress from visitor interaction. This includes well-designed viewing areas, educational signage, and guidelines for visitor behavior.

Enrichment and Stimulus

- *Challenge:* Captive animals may face a lack of mental and physical stimulation, leading to boredom and stereotypic behaviors.

- *Risk Mitigation:* Zoos prioritize enrichment programs, offering various stimuli such as toys, puzzles, and activities to engage animals mentally and physically.

Conservation Messaging

- *Challenge:* Balancing entertainment and education can be challenging. Zoos must convey conservation messages without trivializing the serious threats faced by wildlife in the wild.

- ***Risk Mitigation:*** Zoos emphasize educational programs, interpretive signage, and guided tours to communicate the importance of conservation and the role zoos play in protecting species.

Ethical Considerations

Challenge: Ethical concerns related to captivity, especially for species with complex needs, may arise.

Risk Mitigation: Zoos adhere to ethical guidelines and continuously reassess their practices. Some zoos focus on species that benefit from captivity, such as those involved in breeding and reintroduction programs.

Zoos are actively addressing these challenges through research, collaboration, and ongoing improvements in animal care practices. The aim is to strike a balance between conservation, education, and the ethical treatment of animals in captivity.

Comparisons to other incidents in American zoos

As the North Carolina Zoo grapples with the heart-wrenching loss of Fenn, the 6-month-old giraffe, it prompts a somber reflection on similar incidents that have unfolded in American zoos. Each incident, a unique chapter in the collective story of captive wildlife, reveals both the challenges and the resilience embedded in the world of zoo conservation.

In Waco, Texas, the whispers of sorrow linger from the mysterious deaths of giraffes Penelope and Zuri at Cameron Park Zoo in October. The cause—a complex dance of muscle trauma arising from exertional rhabdomyolysis, a condition echoing the fragility of life within the zoo confines.

Meanwhile, across the nation, a different tale unfolded at the Cincinnati Zoo in 2016. The tragic incident involving Harambe, a silverback gorilla, thrust the challenges of zoo safety into the spotlight. A child's entry into the enclosure resulted in the heartbreaking decision to euthanize

Harambe, underscoring the delicate balance between human curiosity and animal welfare.

The incident with Fenn in North Carolina, much like these narratives, raises questions about the intricacies of captive environments. The collision that led to Fenn's untimely demise invites contemplation on enclosure design, animal behavior, and the perpetual pursuit of safety within zoo landscapes.

In reflecting on these incidents, a common thread emerges—the profound impact on zoo staff. The bonds formed between caretakers and their charges transcend the roles defined by enclosures. The grief counselor summoned in the wake of Fenn's tragedy echoes similar support structures employed to aid the healing process in the aftermath of other incidents.

Yet, within the shadows of these narratives, there are stories of resilience. Zoos, in response to each incident, embark on a journey of introspection and improvement. From enhanced safety measures to refined animal management practices, these stories become catalysts for change.

In this collective narrative of joy and sorrow, challenges and triumphs, American zoos continue to evolve. Each incident becomes a poignant chapter, not just in the story of individual animals, but in the ongoing saga of wildlife conservation. The echoes of tragedy serve as a call to action—a reminder that the delicate dance within zoo environments demands unwavering commitment to the well-being of the creatures under human care.

The responsibility of zoos in ensuring the well-being of their inhabitants

Within the carefully curated landscapes of zoos, a profound responsibility unfolds—a commitment to safeguard the well-being of the diverse inhabitants under human care. The role of zoos extends far beyond the confines of enclosures; it is a custodianship that requires an unwavering dedication to the physical, mental, and emotional health of the resident animals.

In the heart of the North Carolina Zoo, this responsibility takes center stage. Each enclosure, meticulously designed to mirror the natural habitats of its inhabitants, reflects the commitment to providing a home that nurtures rather than constrains. But what does this responsibility truly entail?

1. Physical Health: A Symphony of Care

From the towering giraffe to the playful primates, the physical health of every resident is a priority. Zoo veterinarians, akin to guardians of a vibrant symphony, conduct regular health check-ups, address ailments, and implement preventative care measures. Specialized diets are crafted, exercise routines are encouraged, and medical interventions become a tapestry of care designed to ensure longevity and vitality.

2. Mental Stimulation: Crafting Narratives of Enrichment

Beyond the observable, the mental well-being of zoo inhabitants emerges as a critical facet of their care. The challenges of captivity, the absence of the vast landscapes they once roamed, demand creative solutions. Zookeepers, the architects of mental enrichment, introduce puzzles, toys, and activities that engage the residents' minds. These stimuli become chapters in the narrative of captive life, fostering a sense of curiosity and preventing the onset of boredom.

3. Social Dynamics: Nurturing Connections

In the social tapestry of zoo life, the importance of meaningful connections cannot be overstated. Zookeepers, with an intimate understanding of each resident's social inclinations, carefully curate groupings. From the majestic lions that thrive in familial prides to solitary creatures that find solace in quietude, the social dynamics within enclosures are a reflection of the innate needs of each individual.

4. Enclosure Design: Balancing Space and Safety

The very canvas on which the lives of zoo inhabitants unfold is the enclosure itself. Balancing the need for space with considerations of safety, zoo architects craft environments that emulate the intricacies of the wild. Continuous evaluations and improvements are made to address the evolving needs of the residents, ensuring that the enclosures serve as sanctuaries rather than confinements.

5. Conservation and Education: Guardians of Knowledge

The responsibility transcends the physical and emotional realms—it extends into the realm of conservation and education. Zoos become ambassadors for their inhabitants, raising awareness about the challenges faced by wildlife in the wild. Through guided tours, interpretive signage, and educational programs, zoos weave a narrative that underscores the importance of biodiversity and the role humans play in its preservation.

In the collective narrative of zoos across the nation, the responsibility for the well-being of inhabitants is a pledge—a pledge to be stewards, storytellers, and advocates. It is a commitment to fostering a world where the vibrant tapestry of life, even within the confines of captivity, continues to thrive.

Conclusion

As the sun dipped below the horizon, casting a warm glow over the North Carolina Zoo, Fenn's journey reached its final chapters—a tale of threads unraveling and weaving anew.

Fenn's story began with the delicate dance of life. His birth, marked by tentative steps on a May day, painted the first strokes of a tapestry infused with resilience. Within the giraffe herd, Fenn's spirit shimmered like a vibrant thread, adapting to the curated canvas of the zoo.

Yet, life's narrative is never without unexpected turns. In a heart-wrenching twist, Fenn, startled by a fellow giraffe, collided with a gate. The symphony of joy now echoed with the poignant notes of heartbreak. The zoo community, once filled with the melodies of Fenn's exuberance, now grappled with the silence left in his wake.

Amidst the shadows of sorrow, Fenn's legacy emerged as a source of inspiration. His story, intricately woven into the hearts of readers, surpassed the confines of the zoo. Fenn became more than a giraffe; he became a symbol of the connections we forge with the creatures that share our planet.

In the quiet aftermath, a call to action reverberated through Fenn's narrative. His challenges within the zoo became a mirror reflecting broader threats to giraffes in the wild. Fenn's journey pleaded for giraffe conservation, urging us to safeguard these majestic creatures and their habitats.

Simultaneously, the call extended to the support of zoos. These curated spaces, with their intricate balance of challenges and triumphs, stood as vital bastions for conservation. The commitment to zoos became a commitment to safeguarding the intricate tapestry of life they nurtured—a pledge to protect and educate.

As we closed the book on Fenn's story, we found ourselves at the crossroads of sorrow and hope. In every page turned, Fenn's journey reminded us of the power stories hold—to inspire change, advocate for the vulnerable, and ensure that the echoes of resilience and hope reverberate through the corridors of zoos and the wild alike.

And so, as the final chapter unfolded, we carried Fenn's legacy forward—a story told, threads unraveled, and woven anew in the intricate tapestry of life.

www.ingramcontent.com/pod-product-compliance
Lightning Source LLC
Chambersburg PA
CBHW062329290526
45794CB00005B/1956